Gunther Loses a Whisker

By Jordan White

LPZ Publishing
629 Granite Road, Clinton, TN 37716
This LPZ Publishing Edition October 2023
Text Copyright © 2023 by Jordan White
Images Copyright © 2023 by Jordan White
All rights reserved, including the right of reproduction in whole or in part in any form.
Manufactured in the United States of America
ISBN 979-8-9891689-0-3

"When I look into the eyes of an animal, I see the great work of an awesome God."

-James Michael Cox

Gunther is a Tiger, which is different than a house cat
He can weigh up to 600 pounds, so where would you find him at?
Tigers are native to Asia, but they are losing their natural habitat
Some species of Tigers are already extinct, and that's a fact

Gunther lives at the Little Ponderosa Zoo and Rescue
Ever since he was a cub, this is where he grew
He is bigger now and notices things more
Using his senses, he gathers information and explores

Senses are essential for Tigers, especially in the wild
Using the senses, information is compiled
Whether it be taste, touch, smell, hearing, or sight
The senses help Tigers stay alive

One of Gunther's keenest senses is his sense of touching and feeling
He can use his paws, tail, fur, and whiskers too
All these tools can be very revealing
It is like Gunther is a detective and his senses collect clues

In this story, we will learn about Gunther's whiskers
Why does he have them? What do they do?
He has a roar that is much louder than whispers
So loud that you can hear him all throughout the zoo

He has whiskers short and small, and some grow long and thin
There are some around his eyes and mouth, and even on his chin
Each whisker is different and has a different skill
But every whisker is important in helping Gunther feel

Each whisker has a type, and if you count them there are five
Let us learn about the types of whiskers that help Gunther survive
The Mystacial (meh-stay-shul) whiskers are located on his snout
When he needs to feel around in the dark, these whiskers help him out

Superciliary (sue-per-sill-ee-airy) whiskers are located above his eyes
They protect his eyes when hunting with a protective blink
Even though these whiskers are the smallest in size
Being small does not mean they are not important, Gunther thinks

Tylotrich (tie-low-trick) whiskers are located throughout his body
These whiskers help him understand the space around
They help him find his way when it is dark or foggy
And tell him where he is located relative to the ground

Mandibular (man-dib-you-ler) whiskers are under his snout
And Genal (gee-nul) whiskers are located on his cheek
They count as one type because they both tell Gunther about
The prey that he may seek

Carpal (car-pull) whiskers are on the front legs near the back
These act as a temperature gauge for surfaces he may track
They tell Gunther if something is hot or cool
If he needs a blanket to keep warm, or to cool off in the pool

Losing these whiskers must hurt when they fall
But Gunther assures us that it does not hurt at all
It is like when we lose some of our hair
We do not notice it, because we have plenty to spare

And just like our hair, whiskers grow back
They are never gone long, in fact
Gunther says that it is normal for Tigers to lose a few every now and then
Losing them allows for new and healthier whiskers to grow back in

Gunther has some of the biggest whiskers his Zookeepers have ever seen
But what does the size of a whisker mean
Some whiskers are big, and others are small
But no matter their size, they all eventually fall

Just like human hair, whiskers are different for every Tiger
Some are darker in color, and others are lighter
Also, like humans, no two Tigers are the exact same
If none of them had differences, that would be a shame

Gunther is here to tell you that differences are what make us unique
We all have differences, and that does not make any of us weak
In fact, differences are what make us strong
And help us all to get along

So, next time you see a Tiger, look for the whiskers he wears
Some have been shed, but most of them are still there
If you want to see Gunther, come out to the Little Ponderosa Zoo
And where you find him, there you will find his friends too

And no matter what size or color you are
Your differences will help you go far
Never forget that there is only one of you
And there is only one Gunther, too

In loving memory of James Michael Cox, founder and director of the Little Ponderosa Zoo and Rescue.

January 13th, 1960 – February 9th, 2021

If you would like to donate to Gunther and the Little Ponderosa Zoo, scan the QR code below or visit us online at: *www.littleponderosazoo.com*